Author Dedication:

*For Alec and Paul, their incredibly supportive families,
and all the other soldiers who sacrifice so very much
for the freedom and liberty the rest of us enjoy.
May God bless each of you and those who love you.*

~

Illustrator Dedication:

*For Orvel H. Cockrel and Harry E. Jarboe,
both of whom served our country during World War II.*

Text copyright © 2017 by Ross H. Mackenzie

Illustrations copyright © 2017 by Marvin Jarboe

Hardback edition © 2017 by Patriot Kids, LLC

Book Design by Don Fibich

Layout by David Carrier

All rights reserved. No part of this book may be reproduced or transmitted in any form or by any means, electronic or mechanical, including photocopy, recording, or by any information storage and retrieval system, without permission in writing from the publisher.

The views presented in these pages are purely those of the author and not necessarily those of the Department of Defense or its components.

Requests for permission to make copies of any part of this book should be mailed to the following address:
Patriot Kids, LLC, P.O. Box 10041, Fleming Island, Florida, 32006.

www.patriot-kids.com

ISBN 978-0-9893420-2-5

Printed in the United States of America.

My Soldier Dad

Written by
Ross H. Mackenzie

Illustrated by
Marvin Jarboe

PATRIOT KIDS
Little Books. Big Ideas.
www.patriot-kids.com

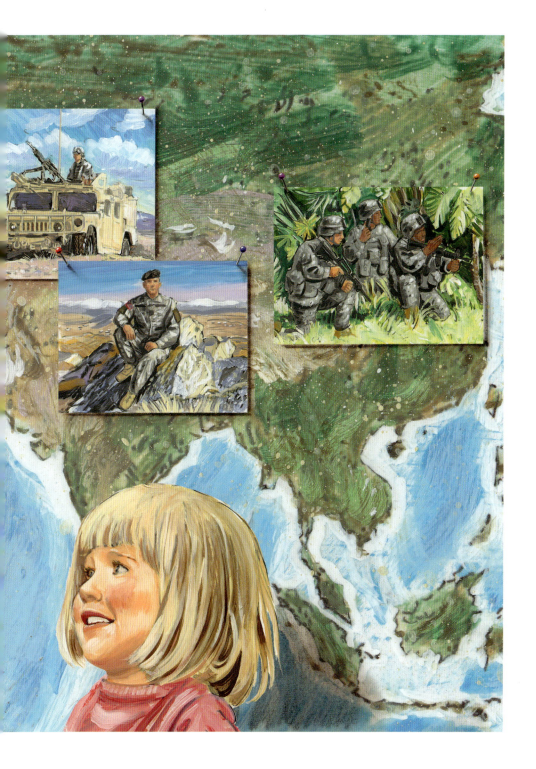

My daddy's a soldier!
He's one of those guys
Who wears a cool uniform
as his disguise.

He goes on deployments
and travels the world,
Defending our freedom –
our great flag unfurled.

When he goes to work,
he wears camouflage clothes.
Those clothes are invisible
clothes to his foes!

Invisible clothes can
help Dad a whole lot:
They help him in deserts
where weather is hot;

They help him in jungles
and mountains and snow;
They help him in places
we don't even know . . .

He tells me some stories
when I go to bed;
His stories are better than
stories I've read!

His stories could go on
forever, I think;
When I listen to them
I don't even blink!

My favorite story of his
that he's told
Occurred in the mountains
where it's very cold . . .

Another great story
he told me one time
Took place in a rainstorm —
in mud and in slime.

Dad's squad was protecting
a headquarters post
From enemy snipers
who move like
green ghosts. . .

The rainforest loomed as
I crawled through the mud –
My squad-mates around me
all covered in crud . . .

We tracked the bad guys
up the sides of two hills –
As silent as snakes
sneaking on window sills . . .

Our rigorous training
paid off on that day;
We captured the snipers
in a peaceful way!

One day Daddy took me
to his soldier base.
It really is quite
a remarkable place!

I saw all the stuff that
he works with each day;
I met a tank gunner;
he even said, "Hey!"

I climbed on a cannon;
I crept in a fort.
When I got inside
I was glad that I'm short!

I saw helicopters
and noisy machines;
I even got grease from
machines on my jeans!

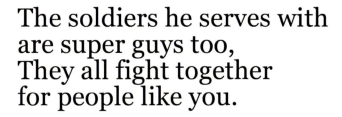

The soldiers he serves with
are super guys too,
They all fight together
for people like you.

Army guys, Navy SEALs,
160th SOAR,
Air Force guys, PJs,
and the Marine Corps . . .

Force RECON and
Delta Force round out the list
(I'm sure there are others
out there that I've missed).

That day was terrific
I spent with my dad!
We had a great time
in the time that we had!

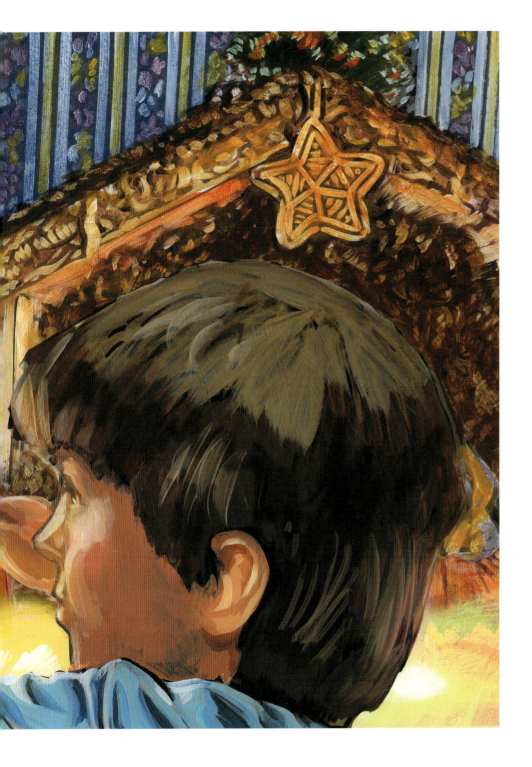

One thing about being
a soldier is this:
My dad goes away a **lot** –
sometimes he'll miss

My birthday and Christmas
and lots of weekends;
So while he's away
I play more with my friends.

I know Daddy's working
when he disappears.
There's no need to worry;
there's no need for tears.

But I still get sad
when he leaves Mom and me.
We're really great buddies,
my daddy and me!

I miss him and miss him,
but boy am I glad
That fighting for freedom's
important to Dad!

Brave men like my dad
risk their lives every day,
Defending us all while
we work and we play.

Across planet Earth
lots of things can go wrong.
And when those things
happen our armies are strong.

Strong armies help countries
stand up for their rights,
And some disagreements
can lead to bad fights.

And if those fights happen
(and sometimes they can),
Our soldiers are ready
to fight – every man!

They bring in their guns and their huge armored tanks;
They bring in **more** soldiers, all standing in ranks.

They bring reinforcements and choppers and planes;
They bring in smart generals with their great big brains.

With everyone working together out there
Our mission's victorious!
That is our prayer.

But soldiers do other
non-fighting jobs too,
They **love** helping people —
it's just what they do!

Some soldiers help folks
who've been hit by a storm;
They help them build shelters
so they can stay warm.

They hand out supplies
if there's been a bad flood
From loud helicopters
that land in the mud.

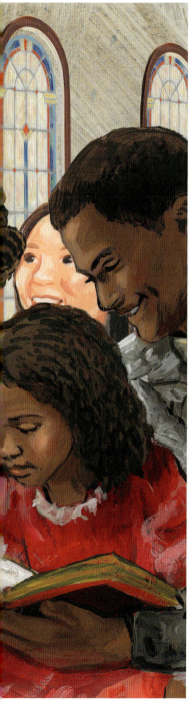

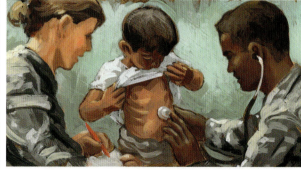

All soldiers help people
in multiple ways —
They work from their hearts
and not for any praise.

They fight for their families,
country, and God.
And nothing could be
more important than God.

Our soldiers are heroes
not seen on the news.
We think of them all
when we pray in church pews.

And Daddy is one of them, I'm proud to say.
I know that he loves me when he goes away.

He's out there protecting us – keeping us free.
I'm glad he's a soldier . . .

Ross H. Mackenzie • Author

Ross H. Mackenzie graduated from the U.S. Naval Academy with a degree in English Literature. He also earned a Master's from St. John's College (Annapolis) and attended Harvard's Kennedy School of Government.

A former Navy pilot, Mr. Mackenzie deployed multiple times on eight different ships to many parts of the world during his 20-year military career. While in the Navy, Mr. Mackenzie taught English literature and writing at the U.S. Naval Academy for four years.

Mr. Mackenzie finds his passions in wide open spaces as an avid cyclist, hiker, and sportsman. He is a devoted father and husband and makes his home in Florida.

Mr. Mackenzie's other books include:
- *Brief Points: An Almanac for Parents and Friends of U.S. Naval Academy Midshipmen* (U. S. Naval Institute Press, 2004).
- *My Sailor Dad* (Patriot Kids, LLC, 2008)
- *Tying Up Water and Other Stories* (Amazon, 2012)
- *My Peaceful Dad* (Amazon, 2012)

Marvin Jarboe • Illustrator

A Kentucky native, Marvin Jarboe joined the Army in 1968 and served as an 81E20 Illustrator. He also earned a Bachelor of Fine Arts degree from Western Kentucky University.

Mr. Jarboe works in advertising, fine art, commercial screen printing, and stained glass. *My Soldier Dad* is his second book with Patriot Kids.